GRADES K-3

Exploring with Color Tiles

Judi Magarian-Gold
and
Sandra Mogensen

Cover Design: Amy A. Berger
Book Design: Jill Wood

Copyright © 1990 by
CUISENAIRE® COMPANY OF AMERICA, INC.
10 Bank Street, Box 5026 White Plains, NY 10602-5026

ISBN 0-938587-17-X

345678910-DAL-98 97 96 95

CONTENTS

BEGINNING EXPLORATIONS   1

   Skills   1

   Materials   1

   Teacher   2

   Partners   6

   Game   8

PATTERNS   9

   Skills   9

   Materials   9

   Teacher   10

   Partners   12

   Game   14

COUNTING 15

   Skills   15

   Materials   15

   Teacher   16

   Partners   20

   Game   22

PLACE VALUE   23

   Skills   23

   Materials   23

   Teacher   24

   Partners   27

   Game   28

NUMBER FACTS 29

Skills 29

Materials 29

Teacher 30

Partners 32

Game 33

SAMPLING AND PROBABILITY 35

Skills 35

Materials 35

Teacher 36

Partners 38

Game 41

ESTIMATION AND MEASUREMENT 43

Skills 43

Materials 43

Teacher 44

Partners 46

Game 50

GEOMETRY 51

Skills 51

Materials 51

Teacher 52

Partners 55

Game 56

BLACKLINE MASTERS 57

INTRODUCTION

NCTM's *Curriculum and Evaluation Standards for School Mathematics* describes a "vision for school mathematics built around five overall curricular goals for students to achieve: learning to value mathematics, becomong confident in one's own ability, becoming a mathematical problem solver, learning to communicate mathematically, and learning to reason mathematicall."[1]

This *vision*, as described in the Standards, reflects our viewpoint of what mathematics is. A successful mahematics program includes expeiences that encourage curiosity and enjoyment while increasing a child's ability to problem solve and communicate mathematically.

The NCTM standards assume the ". . . K-4 curriculum should actively involve children in doing mathematics . . . by interactiong with the physical world, materials and other children."[2] This assumption, that the learniong of mathematics *is* an active process, underlies *Exploring with Color Tiles* (Grades K-3). The classroom enmvironment needs to be one that ecnourages children to explore mathematicla ideas individually, in small cooperative groups, and as a whole class. It should be one that encourages the use of concrete materials when students are exploring, developing and gaining understanding of new concepts.

Exploring with Color Tiles (Grades K-3) is divided into eight units: Beginning Explorations, Patterns, Counting, PLace Value, Nujmber Facts, Sampling and Probability, Estimation and Measurement, and Geometry.

Each unit consists of:

1. Resource Pages. This includes a listing of the Skills covered and the Materials needed for the unit, a Teacher Introduction/Presentation, and a Partners/Cooperative Groups section. The Teacher Introduction/Presentation outlines one or more teacher-directed activities to be used to introduce a particular concept. An overhead transparency, in the form of a blackline master to be reproduced on acetate, is provided to help the teacher model each activity. The Partners/Cooperative Groups section provides independent activities intended to give children the exploration time they need to internalize the unit's concepts. All activities encourage verbal interaction among students, often require

1. Curriculum and Evaluation Standards for Schiool Mathematics. National Association of Teachers of Mathematics, March, 1989. P. 16.
2. Ibid, p. 17.

written summaries, and should be followed by whole class discussion.

2. Worksheet(s). Designed for students working individually or in pairs, they are to be used either during the teacher-directed lesson or as a follow-up investigation. When used independently, post the worsheets and follow with a class discussion so students can share the thinking they did iun completeing the worksheet.

3. Game. Requiring 2 to 4 players, each game is a reinforcement of the concept(s) presented in the unit. Where appropriate, a reproducible gamesheet is provided.

All activities require one-inch square Color Tiles. Cuisenaire's Color Tiles come in sets of 400 in 4 colors — 100 red, 100 yellow, 100 blue and 100 green. Two boxes of 400 each are ideal, although one box will be enough for most activities.

Most units also require a set of Color Tiles for the Overhead Projector, Probability Spinners with Color Divisions (red, yellow, blue, green), dice, one-inch squared paper, butcher (Kraft) paper, crayons, and marking pens.

We hope that your students find the activities rewarding. Our vision is that students become mathematically literate, increase their reasoning and problem-solving skills, learn to work cooperatively with one another, and believe that math *is* fun!

SANDRA MOGENSEN

JUDI MAGARIAN-GOLD

1 Beginning Explorations

Skills

Comparing   Counting
Classifying   Matching
Visualizing   Problem Solving

Materials

Color Tiles, 25-30 per Student   Overhead Pens
Color Spinner   Overhead Projector
Crayons   Dog Cover Gamesheet (p. 00)
Dice   One-inch Squared Paper, 8½" x 11"
Large Pieces of Butcher (Kraft)   (p. 00)
  Paper   Transparency 1 (p. 00)
Overhead Color Tiles   Worksheet 1 (p. 00)

TEACHER
Introduction/Presentation

FREE EXPLORATION

Give children time to explore with the Color Tiles. It is important for children to become familiar with a material before being asked to focus on a specific activity. Free exploration may continue over several classroom periods.

Following are some ways children enjoy using Color Tiles, either on a desk or on the floor:

- Making numbers and alphabet letters
- Building roads and fences
- Designing their own initials and names
- Making animals, flowers, robots, trees, cars, rockets. houses, etc.
- Making symmetrical and asymmetrical designs

Throughout the school year, have children use Worksheet 1 whenever they wish. You can decide how many tiles of each color they are to use, or let them decide.

COPYING

Distribute Color Tiles and one-inch squared paper to students. Have them work in pairs or individually. Using Transparency 1, introduce the following activities.

Make a design with two colors and less than 10 tiles on the overhead projector. Ask students to copy it.

Allow time for children to copy the design, then ask: "Where did you start to copy? What did you do next?" This gives children a chance to develop language skills and math vocabulary.

Make another design, this time using three colors. Ask students to copy it, and again ask them to describe what they did.

Vary this activity by changing the number and color of the tiles used. Extend the activity by having children make designs for each other to copy.

COPY FROM MEMORY

Turn off the overhead projector and build a simple design of six tiles or less. Explain that you will turn the projector on and give the class 15 seconds to look at the design before turning the projector off. They will then be asked to build the design from memory.

Once everyone has made the design from memory, turn the projector on so they can compare their design to yours.

Repeat this activity, increasing the number and color of tiles used.

Allow interested students to build a design on the overhead for the others to copy.

PROBLEM SOLVING

The following problems give children the opportunity to develop listening skills, learn vocabulary and better their visualization skills. Present any or all of them. For each problem, allow children time to answer, then show, or have students show, the solution on the overhead. Ask volunteers to explain how they approached the problem. Discuss the vocabulary used.

"Take 8 red tiles and 1 blue tile. Make a square using these tiles. Put the blue tile in the center."

R	R	R
R	B	R
R	R	R

"Using the same tiles, make a square with the blue tile in the upper, right corner."

R	R	B
R	R	R
R	R	R

"This time, when making the square, put the blue square in the middle of the bottom row."

R	R	R
R	R	R
R	B	R

"Make a square and put the blue tile at the top of the first column."

B	R	R
R	R	R
R	R	R

"Make a checkerboard square using 9 tiles and any 2 colors."

G	Y	G
Y	G	Y
G	Y	G

"Make a checkerboard square using 16 tiles and any 2 colors."

R	Y	R	Y
Y	R	Y	R
R	Y	R	Y
Y	R	Y	R

"Make a square with 9 tiles, 3 tiles of each of 3 colors, so that your square will have columns, each a different color."

R	G	Y
R	G	Y
R	G	Y

"Make a square of 16 tiles, 4 tiles of each color, so that your square will have 4 rows, each a different color."

R	R	R	R
G	G	G	G
Y	Y	Y	Y
B	B	B	B

"This time make a square with 16 tiles, using only red and green tiles. The 4 tiles in the inside must all be the same color."

R	R	R	R
R	G	G	R
R	G	G	R
R	R	R	R

"Use the same 16 tiles to make another square. This time, the 4 corners must all be the same color."

G	R	R	G
R	R	R	R
R	R	R	R
G	R	R	G

"Make a square with 16 tiles using 2 colors. The diagonals must be the same color and the rest of the square the other color."

R	B	B	R
B	R	R	B
B	R	R	B
R	B	B	R

"Take 3 tiles of each of 3 colors and arrange them into a square so that there is exactly one of each color in each row and in each column."

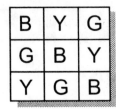

"Use 3 colors. Make a square with 16 tiles. Compare your square to your neighbor's. How are they alike? How are they different?"

"Use 4 colors. Make any size square you choose. Compare your square to your neighbor's. How are they alike? How are they different?"

Note: Activities like these can be offered on task cards at a learning center with an increase in the number of tiles used. For example, squares of 25 tiles or 36 tiles or even 100 tiles could be investigated.

PARTNERS

Cooperative Groups

COLOR TILE MURALS

Give each group a large piece of butcher paper, a supply of Color Tiles, and crayons.

Explain that each group is to make a mural using Color Tiles. As a group, they will need to decide on a theme. They will have to decide what to put on their mural, where everything will go, and who will do what. Once these decisions are made, they will place their Color Tiles on the butcher paper, trace them and color in the outlines.

You may want to elicit theme ideas from the children or suggest your own, such as The Zoo, The Circus, Our School, or Outer Space. Some teachers prefer to set up the butcher paper as shown.

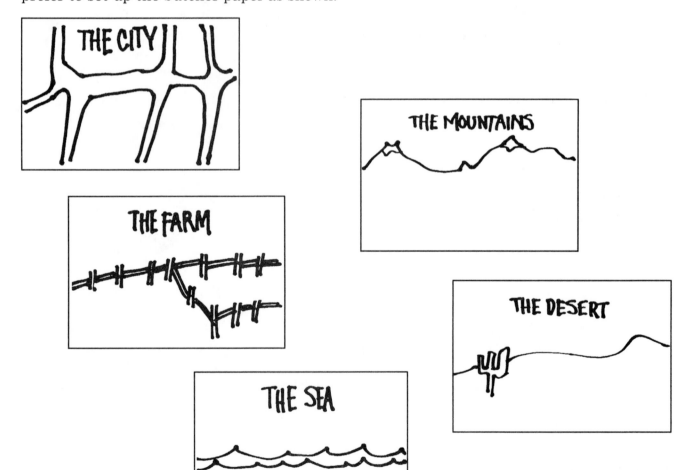

Exploring with Color Tiles © 1994 Cuisenaire Co. of America, Inc.

On a separate sheet of paper each group is to record what pictures are on their mural, the number of tiles used for each picture, and the total number of tiles used for the entire mural.

NAMES Jeffrey   Owen

Gail   Alicia

THE FARM

We made these pictures:

1. Barn   30 tiles
2. Farmer   16 tiles
3. 2 pigs   21 tiles
4. Horse   20 tiles
5. Tree   28 tiles
6. Cow   19 tiles

We used 134 tiles altogether on our mural.

In addition, have each group report, either orally or in writing, how they decided:

- What pictures to put on their mural
- Where to place the pictures
- Who was to do what
- How many tiles they used

GAME

Number of Players: 2-4

Materials:

Color Tiles, at least 100
Dog Cover Gamessheet, one per player
Color Spinner
One Die

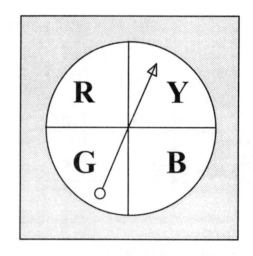

Object:

To be the first player to completely cover your dog.

How to Play:

1. Take turns spinning the spinner. The person who gets green goes first.
2. On your turn, spin the spinner and roll the die. Take the number of tiles that match the die in the color that matches the spinner. (For example, if your spinner lands on red, and you roll a 3, take 3 red tiles.)
3. Put the tiles you just picked on anmy matching squares in your dog. If there is no available space, put the Color Tiles back and wait for your next turn.
4. The winner is the first person to completely cover the dog.

Variation:

Before you begin, cover your dog with the matching Color Tiles. Play as above, but remove tiles when you spin a color and roll the die. The winner is the first player to completely *uncover* the dog.

   Exploring with Color Tiles © 1994 Cuisenaire Co. of America, Inc.

Patterns

Skills

Recognizing Patterns
Sequencing Patterns
Matching Patterns

Counting
Using Ordinal Numbers

Materials

Color Tiles, 25-30 per Student
Color Spinner
Crayons
Overhead Color Tiles
Overhead Pens

Overhead Projector
Transparency 2 (p. 00)
Worksheets 2, 3, 4 (pp. 00-00)
One-inch Squared Paper, 8 ½" x 11"
    (p. 00)

TEACHER

Introduction/Presentation

Distribute Color Tiles and Worksheet 2 to students. Place Transparency 2 on the overhead projector.

Explain to students that in this activity, you will start a color pattern and they are to guess what comes next in order to continue the same pattern.

Place a green tile in the first square of the first row, a yellow tile in the second square, a green tile in the third square, and a yellow tile in the fourth square. Direct the children to copy this pattern and then finish it to the end of the row.

While students complete the pattern at their desks, you complete it on the overhead.

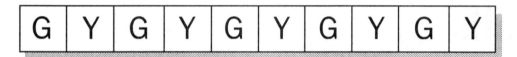

Discuss the pattern you and the class have just completed, Have the class chant the colors in unison. Have them clap hands and snap fingers to correlate rhythmic patterns with sight and hearing. Have them describe this pattern in terms of the alphabet (a, b, a, b, a, b, . . .).

Ask questions like:

  ⊛ How many tiles are needed before this pattern repeats?
  ⊛ How many colors are used?
  ⊛ How many times does this pattern repeat?
  ⊛ What else do you notice about this pattern?
  ⊛ What color is the third tile?
  ⊛ What color is the eighth tile? the fifth tile?

Do several more two-color patterns, each time summarizing with the questions suggested.

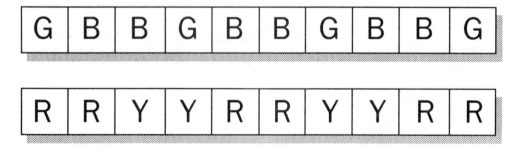

Then work with three-color patterns. Place a blue tile on the first square, a red tile on the second square, a yellow tile on the third square, and a blue tile on the fourth square. Direct students to copy what you have done and finish the pattern to the end of the row.

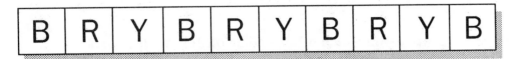

Continue as you did fro the two-color patterns — chanting the colors, clapping hands, snapping fingers and stamping feet, describing alphabetically, and asking the suggested questions.

Do several more three-color patterns.

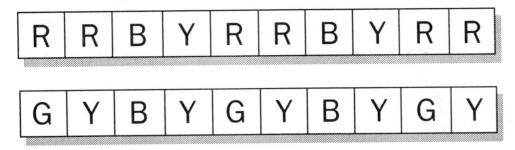

Continue with four-color patterns.

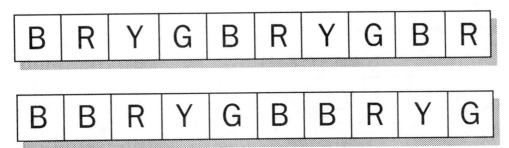

To extend working with patterns, show a two-, a three- and a four-color pattern (or several two-color patterns) on the overhead simultaneously. Clap and snap (clap, snap, stamp, etc.) one of them. Ask students to guess which color pattern you are "acting out" by showing it with their tiles.

Have each student make three or more different patterns. Then, working in pairs, ask them to do what you just modeled. One person will show his/her patterns to the other and "act out" one of them. The other person will guess the pattern. Have students record their patterns on the sheet of one-inch squared paper or on Worksheet 2. Save them and use as task cards throughout the year.

Use Worksheets 3 and 4 to have students investigate patterns independently.

PARTNERS

Cooperative Groups

In small groups, have students extend each other's color patterns. One student in the group starts a color pattern using one or more tiles. Then each group member copies it and adds one more tile to continue the pattern. For example,

1st person

2nd person

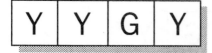

3rd person

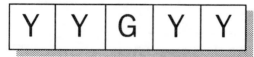

4th person

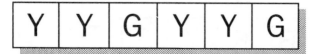

Ask them to summarize what happened as the pattern was extended. Ask the groups to continue until everyone has a chance to initiate a pattern.

Ask pairs on students to make a color pattern that repeats three times. Tell them to record their pattern, Then, using the same tiles, make as many different color patterns as they can, recording each one.

Original pattern:

Exploring with Color Tiles © 1994 Cuisenaire Co. of America, Inc.

Other arrangements:

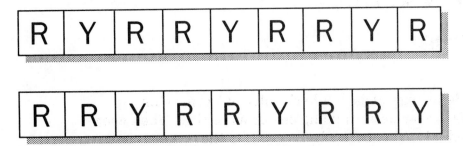

Working in pairs, one partner makes a color pattern of 2, 3, or 4 colors. Then the other partner tries to make that original pattern in reverse.

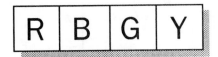

ORIGINAL

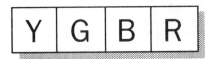

OPPOSITE

GAME

Number of Players: 2-6

Materials:

Color Tiles, 20-30 for each player
Color Spinner
Timer
Objects such as books to use as barriers.

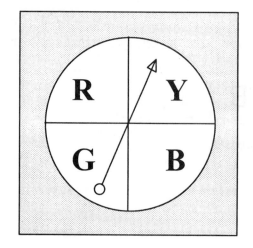

Object:

To be the first player to earn 5 points by building patterns that no one duplicates.

How to Play:

1. Take turns spinning the spinner and setting the timer. The player with the most letters in his/her name begins.

2. On your turn, spin the spinner twice to determine what colors of tiles to use. Announce the colors and ask everyone to build a barrier so that no one can see anyone else's workspace. Set the timer for 60 seconds and say, "Begin."

3. Using the selected colors, everyone makes a pattern that repeats at least twice.

4. When the timer rings, remove the barriers and compare patterns. Each player who has not duplicated a pattern earns one point. The first person to earn 5 points is the winner.

Variation: Use 3 colors. Use 4 colors.

3 Counting

Skills

Counting
Sorting
Matching & Comparing
   Quantities, Sets & Numbers

Reading & Writing Numbers
Exploring Even, Odd & Ordinal
   Numbers

Materials

Crayons
Color Tiles, 25-30 per Student
Large Pieces of Butcher Paper
Number Spinner
Overhead Color Tiles
Overhead Pens
Overhead Projector

Small Paper Bags, Lunch Size
Hundred Chart (p. 00)
One-inch Squared Paper, 8 ½" x 11"
   (p. 00)
Staircase Race Gamesheet (p. 00)
Transparencies 3, 4 (pp. 00-00)
Worksheet 5 (p. 00)

TEACHER
Introduction/Presentation

COUNTING WITH TILES

The following four activities require the overhead projector to be used without a transparency. Students will need Color Tiles.

1 Distribute Color Tiles to students. Place 5 tiles of the same color in a row on the overhead. Ask students to do the same at their desks. As you cover or point to each tile, instruct them to also cover or point to a tile. Cover up the first tile. Count together — 1 said loudly, followed by 2, 3, 4, and 5 all said softly, as you point to each remaining tile. Now cover the first two tiles and count aloud -- 2 said loudly, followed by 3, 4, and 5 said softly. Continue in this manner covering 3 tiles, 4 tiles and finally 5 tiles.

2 Instruct students to make a row of any 8 Color Tiles as you do the same on the overhead. Count the tiles together, pointing to each one as the number is said. Then count backwards from 8 to 1.

Point to the sixth tile and count backwards to 1. Ask students for a number less than 9. Start at the number they choose and count backwards with them. Do this until they have counted backwards from every number possible with 8 tiles. The number 8 is arbitrary. Start with any number that is appropriate for your students.

3 Put a handful of Overhead Color Tiles on the overhead. Sort by color.

Ask questions like these:

- How many red tiles are there? blue? green? yellow?
- How many tiles all together?
- How many red and green tiles together?
- Are there more blue or more green tiles?
- Are there more yellow or more red tiles?
- Which groups have the same number of tiles?
- Which group has the most tiles?
- Which group has the least number of tiles?

Extend this activity with Worksheet 5.

Exploring with Color Tiles © 1994 Cuisenaire Co. of America, Inc.

4 Place a handful of tiles on the overhead. Ask students to count how many tiles there are. Hide 6 tiles (or any amount you choose) under a piece of paper and ask students if there are more or less tiles showing than hidden under the paper. Ask why they think what they do. Show all the tiles again. Hide 1 tile under the paper and again ask if there are more or less showing than hidden under the paper and why. Continue hiding different amounts of tiles and asking more or less are showing. Each time ask students to explain their reasoning.

COLOR TILE AIRPLANES

The following five activities require Transparency 3. Students will need Color Tiles and one-inch squared paper.

1 Ask students to estimate how many tiles are needed to cover the airplane. Cover and count. Have students build the airplane using all four colors. After all have had a chance to build their airplanes, ask for a volunteer to report how many tiles of each color he/she used. Record the student's response on the overhead. Ask if anyone used different amounts and record the response. Continue until you have recorded all the combinations used by the students. Ask if there is still another way to build the airplane.

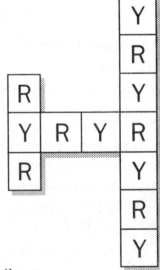

2 Ask students to make the airplane again but this time use only red and yellow tiles and do it so that no two tiles of the same color touch. Follow with questions such as: *"How many yellow tiles did you use? How many red? How many altogether? Can this be done in more than one way?"*

3 Ask Students to make an airplane that uses the same number of each color of tile. Ask students to describe their airplane.

4 Cover Transparency 3 so that the airplane's wings are green, the body is blue, the tail is red and yellow. Ask students the following:

- *How many tiles were used altogether?*
- *Which part of the airplane has more of the same color?*
- *Which color is used the least? How to you know?*
- *Which color is used the most? How do you know?*
- *How many tiles were used altogether for the wings and the body? for the tail and the wings? for the body and the tail?*

5 Have students make an airplane, any size, shape or color(s). Have them record their plane on one-inch squared paper and then color it to match what they built. Display the finished drawings and ask children to compare them. Ask: *"What's the same? What's different?"* Have them sort them into groups, i.e., those with red tails, those with the same number of blue tiles, etc. Ask: *"Can you think of still another way to sort the airplanes?"*

TILES ON THE HUNDRED CHART

Give a pair of children Color Tiles and a Hundred Chart made from the included blackline or a laminated one available from Cuisenaire. For younger children you may want to use only a part of the chart. You will need Transparency 4.

1 Explain that as you count 1, 2, 3, 4, 5, 1, 2, 3, 4, 5, 1, 2, . . . , they are to point to the appropriate square on their Hundred Chart. Each time you say "5", they are to cover the number they are pointing to with a tile. While the students do this at their desks, you will need to do the same on the overhead. When you reach 100, discuss the pattern(s) that can be seen. Now have students remove each tile in order from left to right, starting at the top of the Hundred Chart. As they uncover a number and say it aloud, make a list on the overhead or chalkboard — 5, 10, 15, 20, 25, . . . , 95, 100. Ask students to describe patterns they hear or see.

Have students continue this activity with a partner and a different number. Explain that they are to pick a number between 1 and 10. Using that number, they are to do what you just did with them. One partner will count (i.e., 1, 2, 3, 1, 2, 3, . . .) and the other partner will point to each square, covering some with tiles, until the last square on the Hundred Chart has been reached. Then they are to discuss the pattern(s) they see and describe them in writing. To continue, one partner will remove the tiles, one at a time, calling out each number. The other partner will record. Together they are to discuss the patterns that emerge.

Exploring with Color Tiles © 1994 Cuisenaire Co. of America, Inc.

2 Ask students to take a handful of tiles, count them, and put them into groups of two. Collect the following data from the students.

Number of tiles picked	Number of tiles left
10	0
5	1
•	•
•	•
•	•
•	•

Define even and odd. Explain that if, after putting the tiles into groups of two, no tiles are left, the number of tiles is even. If one tile is left, the number of tiles is odd.

Help students identify each number in the chart as even or odd.

Ask student to cover all odd numbers on the Hundred Chart with yellow tiles. Ask them how many tiles they used. Ask them to describe the patterns they see. Ask them how many numbers are uncovered and to read the numbers aloud. Ask them to discuss the pattern(s) they hear or see.

3 Have student cover the numbers 1 through 10 with blue tiles, 11 through 20 with red tiles, 21 through 30 with yellow tiles and 31 through 40 with green tiles. As you ask questions do the following, have students remove the tile that reveals the answer to each question:

- In the first row, the blue row, take off the first tile. What number is showing?
- In the second row, the red row, take off the third tile. What number is showing?
- Look at the third row. Remove the sixth tile. What number do you see?
- In the fourth row remove the ninth tile. What number do you see?

Continue asking students to remove tiles in this manner. Ask students to create the questions.

PARTNERS

Cooperative Groups

GREATER THAN (LESS THAN)

Ask each pair of students to fill a bag with 25 tiles of the same color. They will also need paper and pencil to record. One person will reach in and pull out some tiles. The other person will do the same. Each one counts how many tiles he/she picked and records the number. Together, the partners decide who picked the most (the least) and circle that number. If they both pick the same amount, they return the tiles and pick again.

Vary the number of tiles in the bag.

Have older children record sentences using > and < signs.

MAKE A STATEMENT

Before beginning this activity review the following words with the children: more, less, same, least, most. Each group will need a paper bag filled with at least 100 Color Tiles. Each student takes a handful of Color Tiles and places them so that all members of the group can see them. Then, taking turns, each person makes a statement about the tiles me/she has picked in relation to what others have picked. The statement must contain one of the five words just reviewed.

For example:

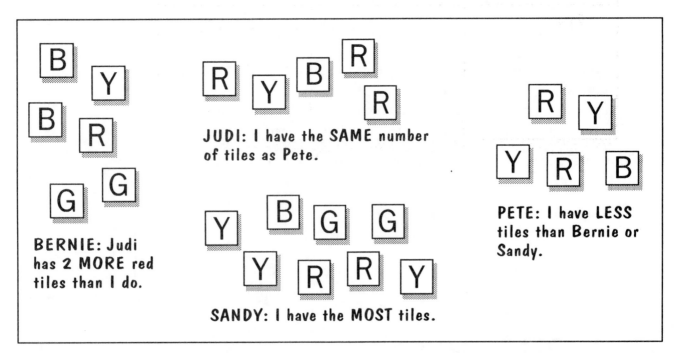

JUDI: I have the SAME number of tiles as Pete.

BERNIE: Judi has 2 MORE red tiles than I do.

SANDY: I have the MOST tiles.

PETE: I have LESS tiles than Bernie or Sandy.

Exploring with Color Tiles © 1994 Cuisenaire Co. of America, Inc.

After everyone has made three statements, give the group a large sheet of butcher paper. Ask the group to draw a picture of each person's tiles, color them, and write one statement about each person's tiles.

Post the students' work. Ask groups to look at what's posted and determine if the statements are true.

MAKE A STATEMENT

Judi		"I have the same number of tiles as Pete."
Bernie		
Sandy		
Pete		

GAME

Number of Players: 2

Materials:

Color Tiles(at least 55)
Staircase Race Gamesheet
Number Spinner

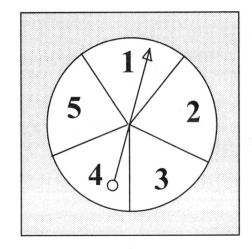

Object: To be the first player to completely fill the staircase 3 times.

How to Play:

1. Player with the most letters in his/her last name begins.
2. Take turns. On your turn, spin the spinner and take the number of tiles showing on the spinner.
3. Beginning at square one, stack the tiles on the gamesheet by putting 1 tile on square one, 2 tiles on square two, 3 tiles on square three, etc.

    For example, suppose you spin a 4. Take 4 tiles, put 1 on square one, 2 on square 2 and 1 on square three. this will complete stairsteps one and two. On your next turn, suppose you spin a 5. You will stack 2 tiles on square three to complete that step and put the remaining 3 tiles on square four.
4. In order to complete the staircase, you must spin the exact number needed.
5. Keep score on the gamesheet. Each time you win a game, put you initials in a circle.

UNIT 4 Place Value

Skills

Identifying Number to 100

Recognizing Patterns on a
   Hundred Chart

Naming Digits in the Ones, Tens,
   & Hundreds Place

Counting

Grouping by Tens

Comparing Two- and Three-digit
   Numbers

Estimating & Finding Area
   & Volume

Materials

Boxes with Lids, each labeled
   A, B, C or D

Color Tiles, 25-30 per Student

One Die

Overhead Color Tiles

Overhead Projector

Overhead Pens

Hundred Chart (p. 00)

Red Tile Wins Gamesheet (p. 00)

Task Cards A through F (p. 00)

Transparency 4 (p. 00)

Worksheet 6 (p. 00)

TEACHER

Introduction/Presentation

In order for children to understand the concept of place value it is important to begin with concrete experiences. Children need to physically make groups of ten objects. They need to count these groups of ten objects. They need to count these groups as if they are one", but must also see each group as being made up of 10 tiles. Being able to count groups instead of individual objects requires new thinking for young children since they are, up to this point, accustomed to counting one number for one object. Also confusing to children is the change in value of a digit as it changes places, i.e., a 2 can be 2 or 20 or 200.

Once modeled, all the activities in this section can be used effectively in learning centers. They all require a Hundred Chart and Color Tiles.

HUNDRED CHART

Give pairs of students a Hundred Chart made from the blackline master (p.00) or a laminated one available form Cuisenaire. Use Transparency 4 to do some or all of the following tasks, as students do them at their seats. This avoids a testing situation by providing children with the opportunity to check their thinking.

1 Ask students to cover all the numbers with a 5 in them with tiles of the same color. Ask how many tiles they used. Direct them to remove the tile covering each number you call out, i.e., 5, 25, 57, 65, 75, etc.

Have students clear their charts and cover all numbers with a 5 in the ones place with a green tile. Ask how many green tiles they used. Instruct students to leave the green tiles on their chart and coverall the numbers that have 5 in the tens place with a blue tile. Ask: *"How many blue tiles did you use? What do you have on 55? Why?"* Repeat this activity focusing on another number between 0 and 9.

2 Ask students to cover the following numbers as you say them: 1, 12, 23, 34, 45, 56, 67, 78, 89, 100.
Ask the following:

- *What kind of pattern do you see?*
- *What is the smallest number you have covered? The largest?*
- *Do you have 23 covered on your board? How do you know?*
- *Which number is 1 more than 33? How do you know?*

- Which number is 10 more than 34? How do you know?
- Put your finger on the tile covering 67. How did you find it? What number is 1 more than 67? Ten more than 67? One less than 67? 10 less than 67?
- What is 5 more than 51? Is that number covered with a tile?
- Beginning at the top left corner, name the number covered by each tile. make a guess, then check by looking under each tile.

3 Ask students to cover the appropriate number with tiles as you give directions like these. For example,

- Cover the number which means one ten and four ones.
- Cover the number which means no tens and seven ones.
- Cover the number which means five tens and no ones.
- Cover the number which has the same number — an 8 — in both the tens and the ones place.

Continue with this type of direction, allowing students time to check answers as you show them on the overhead.

4 Ask students to cover all the numbers that end with the number 6. Ask: "How many squares did you cover? What is the smallest number you covered? the largest number?" Continue by asking students to say each number and then check it by looking under the tile. Ask them to discuss with each other any patterns they see in the column that starts with 6 and ends with 96.

Do the same type of activity, asking students to cover numbers that begin with a 6; that have a difference of two between their digits; etc.

Again, encourage students to discuss the patterns they see.

5 Present number riddles. Have students cover the answers with tiles. Discuss the answers.

- I am thinking of a number with a 6 in the tens place and a 1 in the ones place. I am thinking of a number with a 1 in the tens place and a 6 in the ones place. Are the numbers the same? How do you know?
- Cover the smallest two-digit number on your hundred board.
- Cover the largest two-digit number on your board.
- Cover a three-digit number.
- What number is larger than 2 tens and 3 ones, but smaller than 2 tens and 5 ones?
- Cover the largest two-digit number with a 3 in the ones place.
- Cover the smallest two-digit number with a 5 in the tens place.
- Cover the two-digit number that has a 7 in the tens place and in the ones place.

6 Ask students to work in pairs to make up number riddles for the class to solve.

Ask students to cover the appropriate numbers with Color Tiles as you give any one of these directions. Discuss the numbers they cover. Ask students to clear their board before giving another direction.

- Cover with green tiles all the one-digit numbers.
- Cover with yellow tiles all the numbers that have the same digit in the tens and ones place.
- Cover with red tiles all the numbers whose tens digit is larger than its ones digit.
- Cover with blue tiles all the numbers whose digit is larger than its tens digit.

7 Put the following sets of clues on an overhead. tell students that covering the given numbers with the indicated colors will reveal a picture.

> Cover with yellow: 24, 25, 26, 34, 36, 44, 46
> Cover with red: 35, 45
> Cover with green: 55, 64, 65, 67, 75, 76, 85

Discuss the results.

Have students make their own mystery pictures. Use them as task cards for others to solve.

These additional sets of clues can be used by students working independently.

Set I:

> Cover with yellow: 47, 48, 83, 84, 85, 86, 87, 88
> Cover with red: 38, 74, 75, 76, 77, 78
> Cover with blue: 65, 66, 67, 68
> Cover with green: 56, 57, 58

Set II:

> Cover with yellow: 13, 14, 15, 16, 17, 23, 27, 32, 33, 37, 38, 42, 43, 47, 48, 52, 58, 62, 63, 67, 68, 73, 77
> Cover with red: 45, 53, 57, 64, 65, 66
> Cover with blue: 34, 36

Set III:

> Cover with red: 4, 17, 24, 35, 37, 57
> Cover with green: 7, 14, 27, 34, 36, 38, 47

PARTNERS

Cooperative/Presentation

The following activities have been adapted from *Mathematics With Manipulatives, Color Tiles*, a videotape in the series *Mathematics With Manipulatives* written by Marilyn Burns and produced by Cuisenaire Company. They can also be used effectively in Unit 7, Estimation and Measurement.

TASK CARDS A THROUGH F

Explain to the students that with a partner they will pick a task card and look at the shape. Together they will guess how many tiles they think are needed to cover the shape and record their estimate. Continue to explain that they will test their thinking by covering the shape with tiles, changing colors after every ten tiles. Ask them to record how many tens they use, how many ones, and how many altogether. Ask them to decide if the total is more or less than their original guess.

Have students design task cards like these. Reproduce them on tagboard, laminate them, and use in learning centers.

BOXES

Students, working in pairs, choose a box, estimate how many tiles it holds, and then fill the box. They will need boxes A, B, C, D, Worksheet 6, and at least 150 Color Tiles. Boxes A and B need to be small enough to hold any number of tiles from 11 through 99 and boxes C and D need to be large enough to hold from 100 to 150 tiles. Directions are on the worksheet.

GAME

Number of players: 2-4

Materials:

Color Tiles
Red Tile Wins Gamesheet, one per player
One Die

Object:

To be the first player to earn a red tile.

How to Play:

1. Take turns. the player who rolls the lowest number goes first.
2. On your turn roll the die an pick that many yellow tiles.
3. Place your tiles on your gamesheet in the yellow column.
4. When you have 10 yellow tiles, trade for a blue tile. When you have 10 blue tiles trade for a green tile. When you have 10 green tiles, trade for a red tile.

Variations:

Omit the red column and play Green Tile Wins.

Use only yellow and blue tiles and play Eight Blues Win.

Use a pair of dice and find the sum to know how many yellow tiles to pick.

5 Number Facts

Skills

Counting
Generating Addition &
   Subtraction Facts

Writing Number Sentences
Finding Sums & Differences
Solving Word Problems

Materials

Calculator (optional)
Color Tiles, 25-30 per Student
Construction Paper, 9" x 12"
Dice
Over head Color Tiles
Overhead Projector

Overhead Pens
One-inch Square Paper, 8 ½" x 11"
   (p. 00)
Number Facts Gamesheet (p. 00)
Transparencies 1, 5 (pp. 00)
Worksheet 7 (p. 00)

TEACHER

Introduction/Presentation

TILE STORIES

Distribute 24 Color Tiles, six of each color, to each pair of students and a piece of construction paper to be used as a work mat.

As you tell a story and model it on the overhead, ask the class to do the same at their desks. For example, place three red tiles on the overhead, describe as follows. *"There are three bugs on the grass. Two yellow bugs come and join them. There are five bugs altogether."*

Ask the students:

- How many red bugs are there?
- How many yellow bugs are there?
- How many bugs altogether?

Reinforce by saying, *"Three red bugs and two yellow bugs make five bugs altogether."*

Another story could be: *"I am placing 6 green balloons in the clown's hand. Now I am placing 4 blue balloons in the clown's hand. The clown has 10 balloons altogether."*

Ask the students:

- How many green balloons are in the clown's hand?
- How many blue balloons are in the clown's ?
- How many balloons altogether?

Reinforce with: *"Six green balloons and four blue balloons make ten balloons altogether."*

Have students make up stories fro the rest of the class.

Make up stories that use subtraction. *"I bought a package of 12 lollipops. Five were yellow, two were green and the rest were red. How many were red?"*

TILES IN A ROW

Place Transparency 1 on the overhead. Distribute Color Tiles and 1-inch squared paper to pairs of students.

In the first row place 5 red tiles, followed by 3 blue tiles, one tile on each square. Ask students to do the same and then count together how many tiles there are. Write the answer in the last space of the row. Reinforce by saying, *"5 red tiles plus 3 blue tiles equals 8 tiles altogether."*

In the second row place 3 yellow tiles, followed by 2 green tiles, one tile on each square. After students have done the same at their seats, ask: *"How many tiles altogether?"* Write the response and reinforce by saying, *"3 yellow tiles plus 2 green tiles equals 5 tiles altogether."*

Continue placing tiles in the third row, the fourth row, and so on, each time asking how many altogether, writing the response, and eliciting the addition sentence from the class.

R	R	R	R	R	B	B	B		8
Y	Y	Y	G	G					5
G	G								

Have students give the directions and state the addition sentence.

TILES IN A BOAT

Distribute Color Tiles and Worksheet 7 to each student. Place Transparency 5 on the overhead.

Explain that you are going to place some tiles on Sailboat A and they are to do the same. Then you will tell them how many tiles are needed altogether and they will use that information to figure out how many tiles to put on Sailboat B. For example, place 5 tiles on Sailboat A. *"Nine tiles are needed altogether, how many tiles must go on Sailboat B? Show you answer."* Together state, *"5 + 4 = 9"* and *"9-5=4."*

Continue making combinations up to 20.

HIDE AWAY

In this activity you will place some tiles on the overhead projector and ask students to figure out what you should hide under your hand to leave a particular quantity. For example, place 9 tiles on the overhead projector. After stating that there are 9 tiles showing, ask: *"How many tiles do I need to hide under my hand if I want only five tiles to show?"*

Cover what they say and count on. Reinforce by stating, *"9-4=5."*

Continue this activity varying the amount you start with and the amount to be left.

PARTNERS

Cooperative Groups

STAIRCASE . . . AND MORE

Students will need Color Tiles and one-inch squared paper. Ask pairs of students to build a six-step staircase of one color. Build one on the overhead. (Transparency 1 may be helpful.) When everyone is ready, add a tile of a different color to each row. Have students describe each row in 2 ways. Model for the first row, i.e., 2 or 1 + 1.

R	Y				
R	R	Y			
R	R	R	Y		
R	R	R	R	Y	
R	R	R	R	R	Y
R	R	R	R	R	R	Y

$2 = 1 + 1$

$3 = 2 + 1$

$4 = 3 + 1$

$5 = 4 + 1$

$6 = 5 + 1$

$7 = 6 + 1$

Have students continue this activity independently. Have them build a six-step staircase and add 2 tiles of a different color to each row. Ask them to color the squares on one-inch squared paper to show what they built. Also, ask them to record the combinations and their sums. Have them do it again, this time adding 3 to each row, 4 to each row, or any number you decide.

WHAT'S THE EQUATION?

Pairs of students will need dice and paper and pencil. In this activity, the students will generate subtraction sentences. One person will roll the dice and pick that many tiles. Then he/she will roll the dice again, this time removing the indicated number of tiles. After figuring what is left, the students write the appropriate number sentence. Instruct students to take turns rolling the dice, creating at least eight subtraction sentences.

Fred	Mary Jane
1.	12 - 5 = 7
2.	12 - 2 = 10
3.	9 - 1 = 8
4.	
5.	
6.	
7.	
8.	

1st Roll   2nd Roll

GAME

Number of Players: 2

Materials:

15 red and 15 green Color Tiles
Yellow and blue Color Tiles, about 100
Paper and pencil for each player
Number Facts Gamesheet
Calculator (optional)

Object:

To be the player whose score is the closest to 99.

How to Play:

1. Cover the even numbers on the gamesheet with rd tiles and odd numbers with green tiles.
2. The person with the longest last name begins.
3. Take turns. On your turn, uncover one even number and one odd number.
      Place the tiles just removed in the Bone Pile, not to be used again.
4. Add your numbers. (You may use the yellow and blue tiles to find the sum.) Record your sum in Box 1 on your scoresheet.
5. Continue taking turns until both of you have completely filled your scoresheets. Add you 7 numbers. Record the sum. The person whose sum is closest to 99 wins.

Follow-up:

What do all the numbers on your scoresheet have in common? How are they alike?

Variation:

Make your own gamesheet with different numbers. Use fewer numbers such as 0-8 or two-digit numbers such as 10-21, repeating them randomly on the grid.

Change the target number from 99 to a smaller number.

UNIT 6 Sampling and Probability

Skills

Predicting
Collecting, Organizing &
   Analysis Data

Sampling
Drawing Conclusions
Making Generalizations

Materials

Color Spinners
Color Tiles, 25-30 per Student
Overhead Color Tiles
Overhead Projector
Overhead Pens

Paper Bags
Transparency 6 (p. 00)
Worksheet 8 (p. 00)
One-inch Squared Paper, 8 ½" x 11"
   (p. 00)

TEACHER

Introduction/Presentation

Put 6 Color Tiles (4 yellow and 2 blue) in a paper bag. Display Transparency 6. Discuss with the class the meaning of the words *probability* and *probable*, using word such as likely, probably, possibility, to be expected, odds, chance, reasonable. Help children use each of these words in a sentence.

As you hold up the paper bag containing the tiles, tell students that the bag contains 6 tiles, some yellow and some blue, and they are going to guess how many of each color there is.

Based on what you just explained, ask students what they know for sure about what's in the bag. Possible student replies are:

"There are 6 tiles."

"Some tiles are yellow."

"Some tiles are blue."

"There are no red tiles."

"There are no green tiles."

Draw 6 tiles in the transparency "bag" and list what they know for sure. Ask students how many of each kind might be in the bag. Record their answers in the Question Box.

If students suggest 6 blues, 6 yellows or 6 blues, 0 yellows, record them. At the end of the lesson, discuss why these combinations are not possible.

Explain that to help them decide what's in he bag, they are going to do some sampling. That means they will pick a tile from the bag without looking, record the color of the tile, and return the tile to the bag; and, they will do this several times.

Choose a student to record each draw on the board with tally marks. As students draw a tile, they call out the color and then return it to he bag. Do 6 draws and stop.

Ask students to use the information on the chalkboard to guess what is probably in the bag. Give students time to share their opinions.

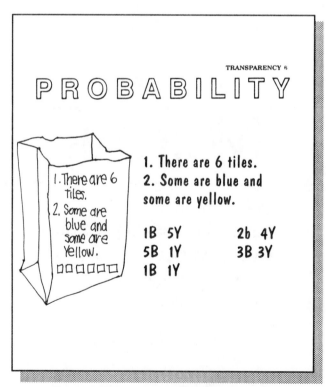

TRANSPARENCY 6

PROBABILITY

1. There are 6 tiles.
2. Some are blue and some are yellow.

1B 5Y	2b 4Y
5B 1Y	3B 3Y
1B 1Y	

Do this two more times, each time drawing and replacing 6 tiles. After each 6 draws have students share their thinking.

Finally, show the tiles in the bag, counting aloud the 4 yellow and 2 blue.

Note: Although 6 Color Tiles are used for this lesson, any number may be chosen.

PARTNERS

Cooperative Groups

HOW MANY OF EACH?

Divide students into small groups. Give each group a paper bag into which you have put 6 tiles (or any amount you wish) of 3 different colors — yellow, blue and green. Make each bag different. For example, make 1 yellow, 1 blue, 4 green; make another 3 yellow, 2 blue, 1 green; another, 2 yellow, 2 blue, 2 green, etc.

Explain to the students that their task is to guess what's in the bag in the same way they guessed what was in the bag in the introductory lesson. *"You will need paper and pencil and someone to do the recording. Take turns picking and replacing tiles. Stop after 6 picks and discuss your results. Pick and replace 6 more times, and again stop and discuss your results. Continue until you think you have drawn enough to make a decision about what's in your bag. Discuss what you think. Then each of you write what you think is in the bag and why."*

When everyone in the group has shared what they wrote, instruct groups to check the contents of their bag. Have each group share with the class what happened in your group.

WHICH BAG?

Fill one bag with 3 red tiles and 7 blue tiles and another with 3 blue tiles and 7 red tiles. Label one bag #1 and the other, #2. Tell students that one bag has 3 red tiles and 7 blue and the other has the opposite, 3 blue and 7 red, but do not tell them which is which. their task is to pick a bag and by sampling decide its contents.

With a partner, students choose one of the bags. They take turns reaching into this bag, pulling out a tile, recording what they see, and returning the tile to the bag. Instruct them to shake the bag between picks. When they have recorded 10 picks, they are to discuss which bag they think they have and why. Instruct them to pick, record, and return a tile ten more times and again discuss which bag they think they have. When they feel "certain" about which bag they have, ask them to report, in writing, what they think and why.

TWO SPINNERS

In this activity, students investigate spinners to determine if certain color combinations come up more or less than others.

Each pair of students needs 2 color spinners, Color Tiles, and one-inch squared paper which they will set up like this:

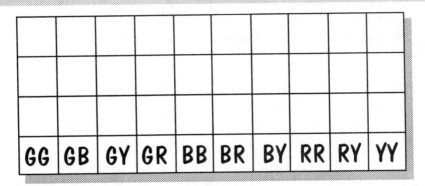

GG	GB	GY	GR	BB	BR	BY	RR	RY	YY

Both spin their spinners at the same time, take a tile of the indicated color and stack them on the appropriate square on the recording sheet.

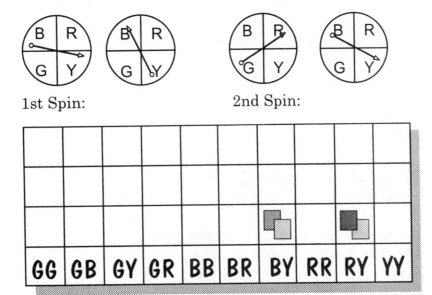

1st Spin: 2nd Spin:

GG	GB	GY	GR	BB	BR	BY	RR	RY	YY

After 10 spins, students check to see if any column seems to have more tiles than the rest. Ask them to spin 10 more times and check again. Ask them to continue spinning until they feel confident that they can make a statement about the chances of any combination coming up more than any other.

Have students share, either verbally or in writing, what they discovered by doing this spinner experiment.

SAME COLOR

Small groups of students investigate how many draws it takes before 2 tiles of the same color are picked. Students will need 8 tiles, 2 of each color, a paper bag, and a group recording sheet.

Instruct the students to put the tiles in the paper bag. As them top predict: *"If you were to reach in the bag and pull out two tiles at the same time, how many tries would it take to pick two tiles that are the same color?"*

Explain that each group member is to draw 2 tiles from the bag without looking, check the color, and return the tiles to the bag until he/she gets a match. they need to keep track of how many reaches into the bag it takes before two of the same color tiles are picked.

Name	Prediction	Picks
Cliff	10	~~IIII~~
Shalamar	11	II
Isaac	14	

After everyone has had a turn, ask groups to look at their data. With this in mind, ask them to do the experiment again, first making a prediction about how many tries it will take to get a match.

In a whole class discussion, have groups report what happened.

LADDER CLIMB

Partners need Worksheet 8, Color Tiles, and a Color Spinner.

Ask students: *"If you spin the spinner 20 times, how many of each color do you predict you will get? Why?"* Have them record their guesses on the worksheet.

Then, instruct them to spin 20 times, each time putting a tile matching in color to the spinner on the ladder. Ask students to count the tiles on the ladder and record how many of each color came up.

Ask students to then clear the ladder and repeat the investigation — predicting, spinning, filling the ladder, counting, and recording.

On the back of the worksheet have students write what they think would happen if they did this experiment another 20 times.

Note: If students spin more colors than the spaces on the ladder, either have them continue adding tiles to the ladder beyond the edge of the paper, or have them climb the ladder again, stacking tiles on top of those already there.

Variation:

Make 1 or 2 spinners with color sections divided into unequal parts, like this:

Have students choose a spinner and predict, spin, fill, and record as before. Ask students to compare these results to their original results. Ask them to compare the two spinners.

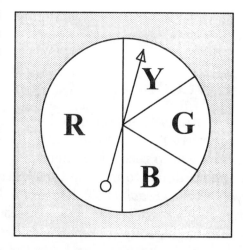

GAME

NO REDS!

Number of Players: 2-4

Materials:

40 Color Tiles, 10 of each color
Paper bag

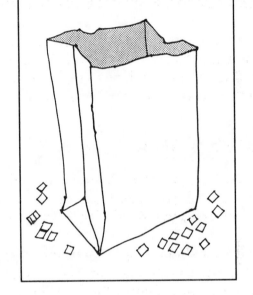

Object: To be the first player to collect 10 tiles.

How to Play:

1. Put all the Color Tiles into the paper bag and shake it.
2. Take turns. On your turn, reach into the bag without looking and take out a tile. If the tile is red, return it to the bag and pass the bag to the next player.
3. If the tile is not red, keep the tile, placing it in front of you. Continue reaching into the bag, picking and keeping tiles until you pick a red one.

Extension:

After you have played the game at least three times, discuss the following:
Is this a fair game?

---

7 Estimation and Measurement

Skills

Comparing
Matching
Predicting Measuring Counting

Investigating Area, Perimeter,
Length, Width, Weight, Volume

Materials

Assorted Containers
Paper
Overhead Color Tiles
Balance Scale
Color Tiles, 25-30 per Student
Crayons
Dice
Inch-rulers (optional)

Large Pieces of Butcher (Kraft)
Overhead Projector
Overhead Pens
Post-it Notes
One-inch Squared paper, 8 ½" x 11"
    (p. 00)
Transparencies 7, 8 (pp. 00-00)
Worksheets 9, 10 (pp. 00-00)

TEACHER
Introduction/Presentation

THE BIG E

Display Transparency 7. Introduce the terms *estimate* and *estimation*, eliciting from students related words or expressions such as guess, judge, "eyeball", "in the ballpark", etc.

Ask students to estimate how many overhead Color Tiles they think it would take to cover the E. Put one overhead tile on the projector to help students think about their estimates.

Record their estimates from the smallest to the largest in he Question Box on the transparency. Use tally marks for numbers given more than once.

Ask students to explain how they determine their estimates. Discuss the smallest and largest estimates, the difference between the two, the number chosen most often, the number chosen least often.

Cover the top row of the E with Color Tiles. Give students the opportunity to change their estimates. As you record the new estimates on the overhead, ask students to explain why they changed them.

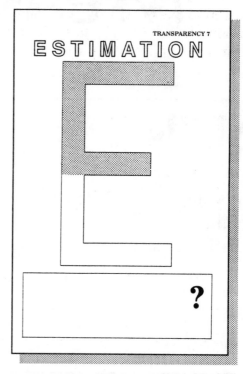

Add tiles to the middle of the E.

Once again, encourage anyone who wants to change their estimate to do so and explain why. record the new numbers.

Fill in the rest of the E and count together the number of tiles used.

Extend this activity with Worksheet 9.

For the following activities, students will need Worksheet 10 and a supply of Color Tiles. You will need Transparency 8.

AREA

Ask students to estimate the area of the Robotile, explaining that this means they need to guess ho many tiles would completely cover the Robotile. Discuss their estimates and then have them cover the Robotile with their tiles.

Have students find the area of other parts of the Robotile.

PERIMETER

Ask students to estimate the perimeter of the Robotile, that is, how many tiles can go around the outside of the Robotile. Have them find the perimeter and explain how they did it. Explain, if students didn't, that the perimeter can also be found by counting all the exposed ("open") sides of each tile in the Robotile.

LENGTH AND WIDTH

Ask students what is meant by the length of the Robotile; what is meant by the width of the Robotile. Together find out how many tiles long and how many tiles wide it is. Have students measure the length and width of other parts of the Robotile — its head, its body, its arms, its legs.

(You might want to distribute inch-rulers and have students do the measuring again, but this time, in inches.

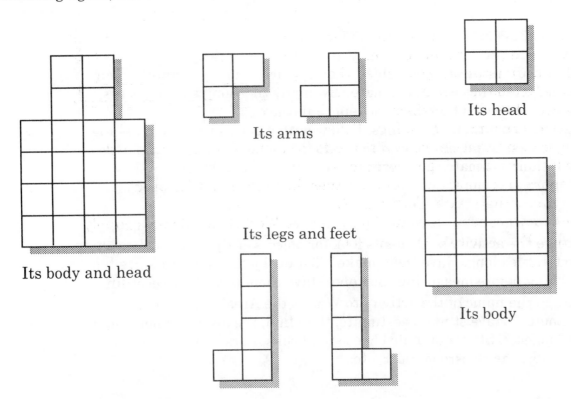

Its arms

Its head

Its body and head

Its legs and feet

Its body

PARTNERS

Cooperative Groups

HOW LONG?

In this activity, pairs or small groups of students work together to measure the tile lengths of objects in the classroom, always estimating first.

Give students a bag of Color Tiles and a large piece of butcher paper to be organized like this

NAMES _____		
_____		
Object	**Estimate**	**Actual**
1.		
2.		
3.		
4.		
5.		

Explain that they are to select 10 objects to measure. As they examine each object they selected they are to estimate how many tiles long each object is, record the estimate, and then find the length in tiles.

Have groups post their recordings. Follow with a class discussion, asking questions such as, "*What did you find that was 2 tiles long? 3 tiles long? 10 tiles long?*" Ask if their estimates got closer to the actual measurement as they measured more and more.

Variations:

- Change the activity so students look for objects of a particular length, like ten tiles long. Have them make a list of objects they think are ten tiles long before measuring each one. Have them post their recordings.
- Change the name of the activity to What's the Area? What's the Perimeter? Have students estimate, then find, the area and perimeter of 10 objects. Students could choose textbooks, library books, desktops, tabletops, the classroom floor, etc.

Exploring with Color Tiles © 1994 Cuisenaire Co. of America, Inc.

Change the name of the activity to What's the Volume? have several small containers available. Two students, working together, guess how many tiles each container holds. then, they fill each container to the top. Once filled, they remove the tiles and put them in groups of ten to count. Have them record and post their findings.

| NAMES | Andy H. |
| | Maria V. |

Container	Holds
Milk container	26
Coffee cup	32

ROWS OF TILES

In this activity students make longer and longer rows of tiles, looking for patterns to help them predict the area and perimeter of a row of any length.

Ask students to find the area and perimeter of

1 tile

2 tiles

3 tiles

. . . up to 10 tiles.

Ask them to record their findings. In a whole class discussion, have students share the patterns they found.

TILES TALL

In a corner of the classroom choose two students to lay down 60 tiles in a line, so that the tiles touch each other. Have them switch color after every 10 tiles.

| R | R | R | R | R | R | R | R | R | R | Y | Y | Y | Y | Y | Y | Y | Y | Y | R | R | ...

In pairs, have students find out how many tiles tall they are. One student lies down next to the tiles with the sole of one foot as close to the first tile as possible. he other student counts to find out approximately how many tiles tall his/her partner is. Once both students have found their tile heights, each records his/her name and the number of tiles on a post-it note. Have students put their post-it notes on the chalkboard in order from the smallest number to the largest.

Have students line up from shortest to tallest. Call off the names according to the post-it notes. Ask: *"Does the order match the way you are lined up? Why or why not?"*

HAND AND FOOT

In this activity students will be comparing their hand and foot tile measurements to those of other students.

Divide students into small groups and distribute blank paper, 8 ½" by 11". Ask group members to help each other trace one "shoeless" foot and one hand o their papers. Ask students to use tiles to find he length, width, area and perimeter of their foot and hand shapes. Explain that "close enough" is okay, since they probably won't be able to get an exact number of tiles to fit. Have them record all their measurements and post every group member's drawings on a large sheet of butcher paper. ask each group to indicate the hand with the largest area, the longest hand, the longest foot, the widest hand, the widest foot, the hand with the largest perimeter, and the foot with the largest perimeter.

After all results are displayed, ask each group o make comparison statements about their findings. For example, *"Jack had the widest hand. It measured 5 tiles across."* or *"Jose had the longest foot. It was 9 tiles long."* Ask questions such as: *"Did the person with the shortest foot also have the smallest area?"*

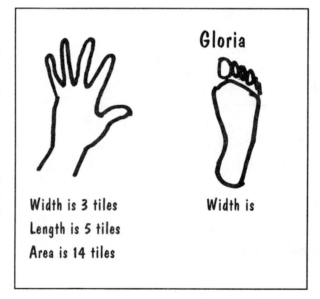

Gloria

Width is 3 tiles
Length is 5 tiles
Area is 14 tiles

Width is

 Exploring with Color Tiles © 1994 Cuisenaire Co. of America, Inc.

DIFFERENT PERIMETERS

Have pairs of students choose 4 tiles of the same color.

Ask them to arrange the tiles into as many different shapes as possible, making sure one side of a tile completely touches another side, like this:

Have students record their shapes and the perimeters on one-inch squared paper. Ask then to find the area of each shape and record that too. Ask them to indicate the shape with the largest perimeter and the one with the smallest perimeter.

This activity can be done again with 5 Color Tiles; with 6 Color Tiles.

In a whole class discussion, ask students to describe the shape with the smallest perimeter; describe the shape with the largest perimeter. Ask: *"Did different shapes have the same perimeter? the same area?"*

SAME PERIMETERS

In this activity, students working together in pairs will use up to 12 Color Tiles to investigate shapes with he same perimeter.

Instruct students to make 5 different shapes whose perimeters are the same. Each tile must touch another completely on at least one side, as in the pervious activity, Different Perimeters. Have them copy the shapes on one-inch squared paper and record the perimeter and the area of each.

Have students describe what they notice about their shapes and their areas and share their thinking with the class.

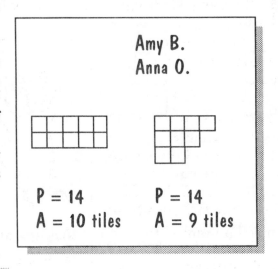

TILE BALANCE

Each pair or small group of students needs a balance scale and 5 Color Tiles on one side of the balance and then find objects in the classroom that weigh about the same. Have hem record and post their findings.

Do the same activity using 10 Color Tiles on one side of the balance.

GAME

FILL IT UP

Number of Players: 2

Materials:

Color Tiles, 36 each of two different colors
A die for each player
Gamesheet, one 6" by 6" square, cut from one-inch squared paper.

Object:

To be the last player to fill a square on the gamesheet

How to play:

1. Select the color you want to use.
2. Take turns. On your turn, roll a die and take that number of tiles. Put them on the gamesheet so that each of your tiles completely touches another tile on at least one side.

    For example, if you roll a 3, you may arrange your tiles in any of these ways:

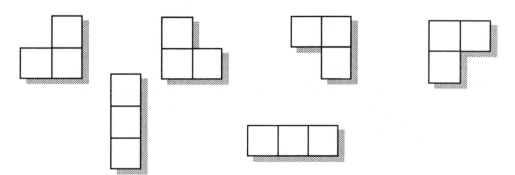

3. If you roll a number that you cannot use, you lose your turn,

Geometry

Skills

Recognizing & Comparing Shapes
Investigating Symmetric Designs
Investigating Geometric Rotations

Adding & Subtracting
Counting
Visualizing

Materials

Color Tiles, 25-30 per Student
Dice
Mirrors
Overhead Color Tiles

Overhead Projector
Overhead Pens
One-inch Squared paper, 8 ½" x 11"
   (p. 00)

TEACHER
Introduction/Presentation

SQUARE OUTLINES

Distribute Color Tiles and one-inch squared paper to students. You will need transparency 1 and Overhead Color Tiles. Ask students to build three squares — the smallest square possible, the next largest square, and the next largest. Build the same three squares on the overhead.

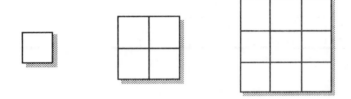

Ask students to compare the squares. *"How are these squares alike? How are they different? Can you predict how many tiles will be needed for the next largest square? Build it."*

Remove as many tiles as possible from each of the squares so that an outline of a square remains. Illustrate by removing the center tile from the 9-tile square. Ask: *"How many tiles would you remove in the 16-tile square in order to leave a square outline? Try it."* Ask students why the 1-tile square and the 4-tile square can't be done.

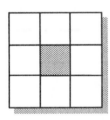

Remove 1 tile

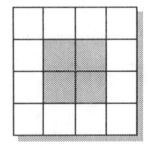

Remove 4 tiles

Instruct students to work in pairs to investigate larger and larger squares. Ask them to predict how many tiles they will need to predict how many tiles they will need to build each square. Then, after building each one, instruct them to remove the largest number of tiles from the center that will leave an "outlined square". To record, have them draw and color what is left on squared paper. Also, have them record the number of tiles they begin with

---

 Exploring with Color Tiles © 1994 Cuisenaire Co. of America, Inc.

and the number of tiles they remove. In a follow-up discussion ask: *"How many tiles did you remove when making a 25-tile square? How many tiles did you remove from a 36-tile square? Can you predict for a 49-tile square? for any size square?"*

FOUR SQUARE

Using four tiles, three of one color and one of another, ask students to make a square. Have students describe what they built. As a student describes the square, build it on the overhead. Continue with students responses until you have built all possible arrangements.

Ask pairs of students to look for relationships among these squares and then share their thinking with the whole class. If students do not notice, show how each square is a rotation of another.

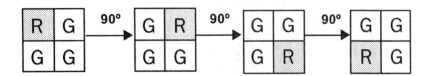

Ask students to make as many squares as possible using 2 tiles of one color and 2 of another color. After giving then time to explore, ask them how the squares can be rotated so one square will look like another.

SYMMETRIC DESIGNS

In addition to Color Tiles and one-inch squared paper, students will need mirrors.

Make the following design on the overhead. Ask each student to copy it.

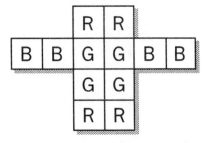

Explain that you are looking for a way to divide the design into two matching pieces. Place a pencil on the figure so that both sides match. Once the class agrees that you have two matching pieces, remove the pencil and draw a line. Explain that this line is as line of symmetry. Ask students to place a mirror on the line of symmetry. *"Does what you see in the mirror match the piece that is showing?"*

Make another design and have students copy it and find its line of symmetry.

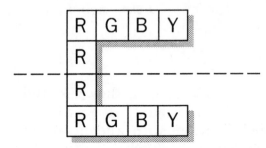

Make a design that has more than one line of symmetry; that has 0 lines of symmetry. Discuss.

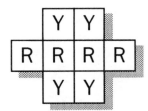

2 lines of symmetry

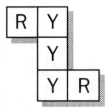

0 lines of symmetry

Have students make a design, find its line(s) of symmetry, and record both the design and the lines of symmetry on one-inch squared paper. have them check each other.

PARTNERS

Cooperative Groups

PENTATILE

In this activity, students working in small groups make shapes using 5 tiles and then investigate which are symmetric and which can fold into a topless box. Each group will need Color Tiles, one-inch squared paper, and the following directions.

Directions:

1. Make a shape with five square tiles.
2. Make sure that at least one full side touches one full side of another.

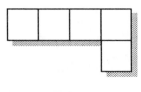
This

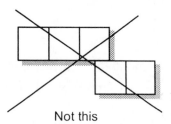
Not this

3. Find all the different shapes that are possible. Record them on squared paper and cut them out.

Ask students to answer the following:

* Which pentatile have symmetry and which ones do not?
* Which pentatiles can fold into a topless box and which one cannot?

Extension:

Have students brainstorm other ways to sort these shapes.

GAME

Number of players: 2-4

Materials:

Color Tiles, about 100
4 Dice
Recording Sheet (shown in example below), one per person

Object:

To be the first player to earn 21 points.

How to Play:

1. Player with the most letters in his/her name begins.
2. Take turns. On your turn roll all 4 dice. Find the total and select that many Color Tiles.
3. Using all the tiles you've picked, make as many different rectangles, including squares, that you can. Record the dimensions of each. (Note: A 4-by-2 rectangle is the same as a 2-by-4. Also, color does not matter.)
4. Record you score. A square is worth 5 points and each rectangle is worth 3 points. For each rectangle your opponent can make that you have not included, you lose 1 point.

   For example: You roll 1, 1, 4, and 2 for a total of 8.
You earn 6 points.

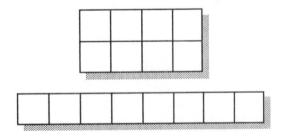

Nina		
Number of tiles	Dimensions	Points Earned
8	2 x 4	3
	1 x 8	3

Blackline Masters

- **Transparencies** (pp. 00-00)

- **Worksheets** (pp. 00-00)

- **Gamesheets** (pp. 00-00)

- **The 100 Chart** (pp. 00-00)

- **Task Cards** (pp. 00-00)

- **One-inch Squared Paper** (pp. 00-00)

1	2	3	4	5	6	7	8	9	10
11	12	13	14	15	16	17	18	19	20
21	22	23	24	25	26	27	28	29	30
31	32	33	34	35	36	37	38	39	40
41	42	43	44	45	46	47	48	49	50
51	52	53	54	55	56	57	58	59	60
61	62	63	64	65	66	67	68	69	70
71	72	73	74	75	76	77	78	79	80
81	82	83	84	85	86	87	88	89	90
91	92	93	94	95	96	97	98	99	100

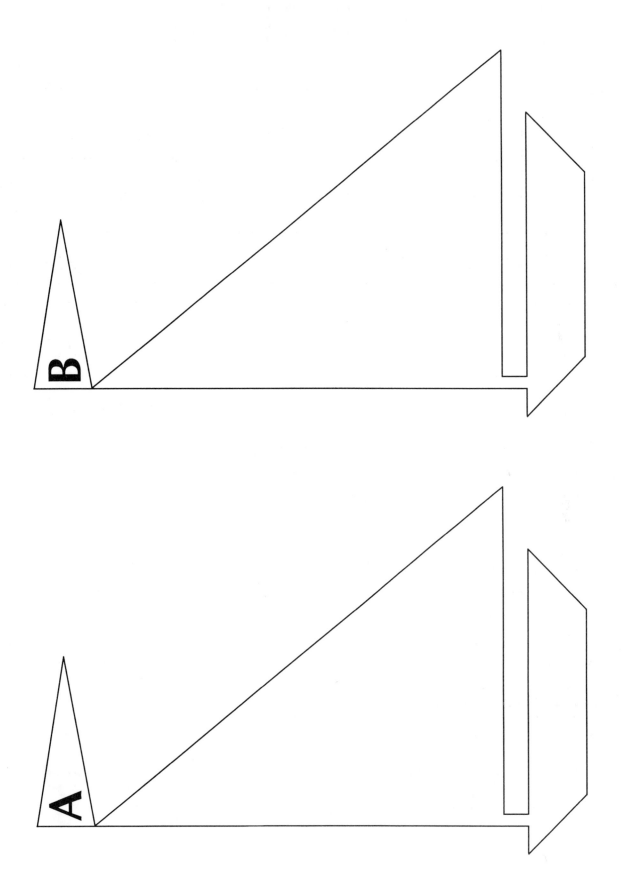

PROBABILITY

?

ESTIMATION

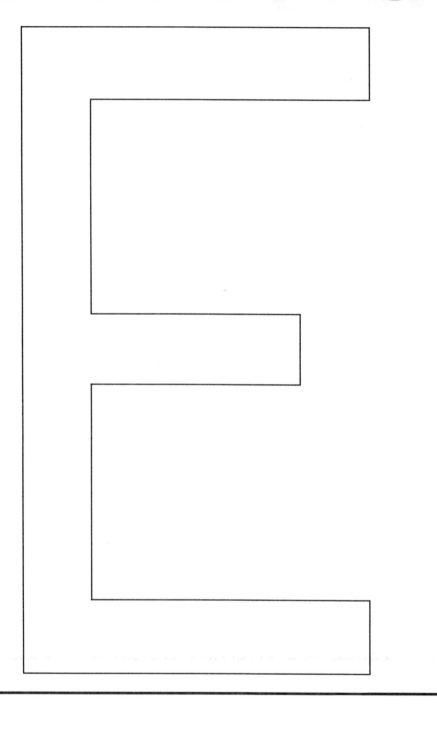

?

ROBOTILE

Exploring with Color Tiles © 1994 Cuisenaire Co. of America, Inc.

NAME _____

Make a design or picture that uses
exactly:

RED	BLUE	GREEN	YELLOW

Color it.

How many tiles did you use altogether? _____

How many empty squares did you leave? _____

NAME _____

Make a ☐-color pattern using tiles.

Color the squares. Answer the questions below.

1. What colors did you use?
2. How many of each color did you use?

   Blue _____ Red _____ Yellow _____ Green _____

3. How manty tiles did you use altogether?
4. If you added 1 more tile to your pattern, what color would it be?
5. If you substitute alphabet letters for each color, what would your pattern look like?

Exploring with Color Tiles © 1994 Cuisenaire Co. of America, Inc.

NAME ⎯⎯⎯⎯⎯

Decide with your partner how many colors and which colors of Color Tiles you
want to use to make a pattern on the worm's body.

Make you pattern, starting at the head and ending at the tail.

Color you worm.

How many times does your pattern repeat? ⎯⎯⎯⎯⎯

Give your worm a name. ⎯⎯⎯⎯⎯

Exploring with Color Tiles © 1994 Cuisenaire Co. of America, Inc.

COLOR TILE SORT

NAME _____   NAME _____

Directions: 1. Work with a partner.
2. Get some color tiles.
3. One of you grabs a handful of color tiles.
4. Together, sort the tiles by color
5. Now, answer the questions below.

A. Do you have more *blue* or *green* tiles? _____

B. Do you have more *yellow* or *red* tiles? _____

C. Do any groups have the same number of tiles? _____
   If so, which colors? _____

D. Which group has the *most* tiles? _____

E. Which group has the *least* tiles? _____

F. How many *red* tiles do you have? _____
   blue tiles? _____   *green* tiles? _____   *yellow* tiles? _____

G. How many tiles did you grab altogether? _____

H. How many *red* and *blue* tiles do you have altogether?

   _____

I. How many *green* and *yellow* tiles do you have altogether?

   _____

Box A

Guess _____

How many? _____

10's	1's

Box B

Guess _____

How many? _____

10's	1's

Box C

Guess _____

How many? _____

10's	1's

Box D

Guess _____

How many? _____

10's	1's

1. Work with a partner. Choose box A, B, C, or D.

2. Guess how many tiles fill your box.

3. Record your guess.

4. Fill your box with as many Color Tiles as possible. Be sure the lid can fit on the box.

5. Count the tiles.

6. Record your answer. Record how many 10's and how many 1's you counted.

7. Choose another box and follow directions 2 through 6.

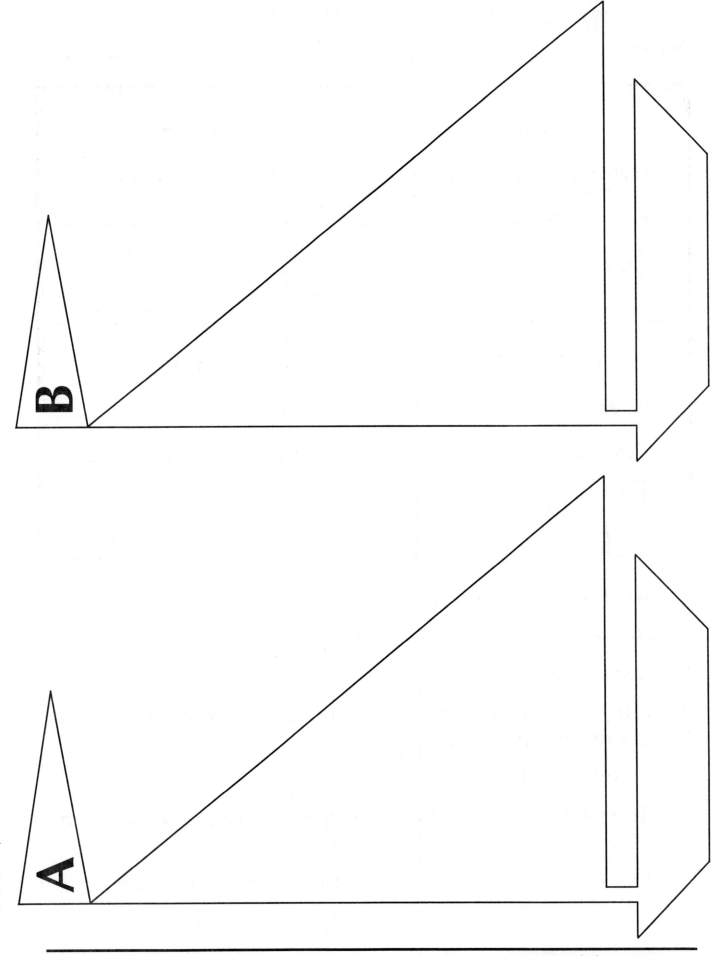

LADDER CLIMB

NAMES: _____

_____

GUESS #1

__ __ __ __
B R Y G

COUNT #1

__ __ __ __
B R Y G

GUESS #2

__ __ __ __
B R Y G

COUNT #2

__ __ __ __
B R Y G

---

25 TILES

NAME _____   NAME _____

1. The rectangle below holds exactly 35 color tiles.

2. Draw a shape inside the rectangle that you think can be covered with exactly 25 tiles.

3. Use 10 tiles of one color, 10 of another, and so on until your shape is filled.

4. Count the tiles.

5. How many? 10's: _____   1's: _____

6. Was your number more or less than 25? _____

7. On the other side of this paper, try this experiment again, making a shape that you think can be covered with 25 tiles.

8. How many tiles? 10's: _____   1's: _____

9. Was your number more or less than 25? _____

ROBOTILE

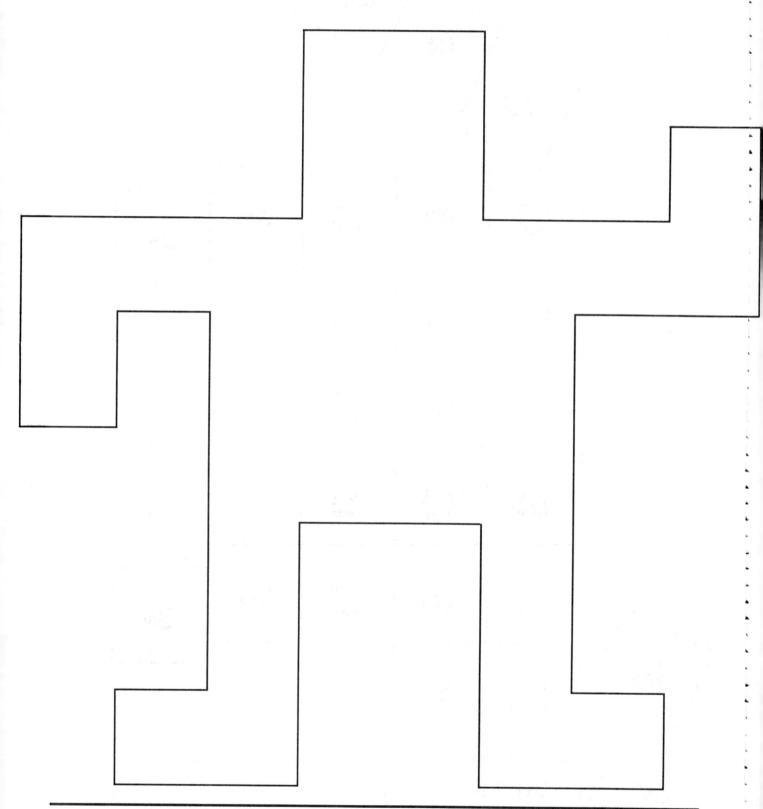

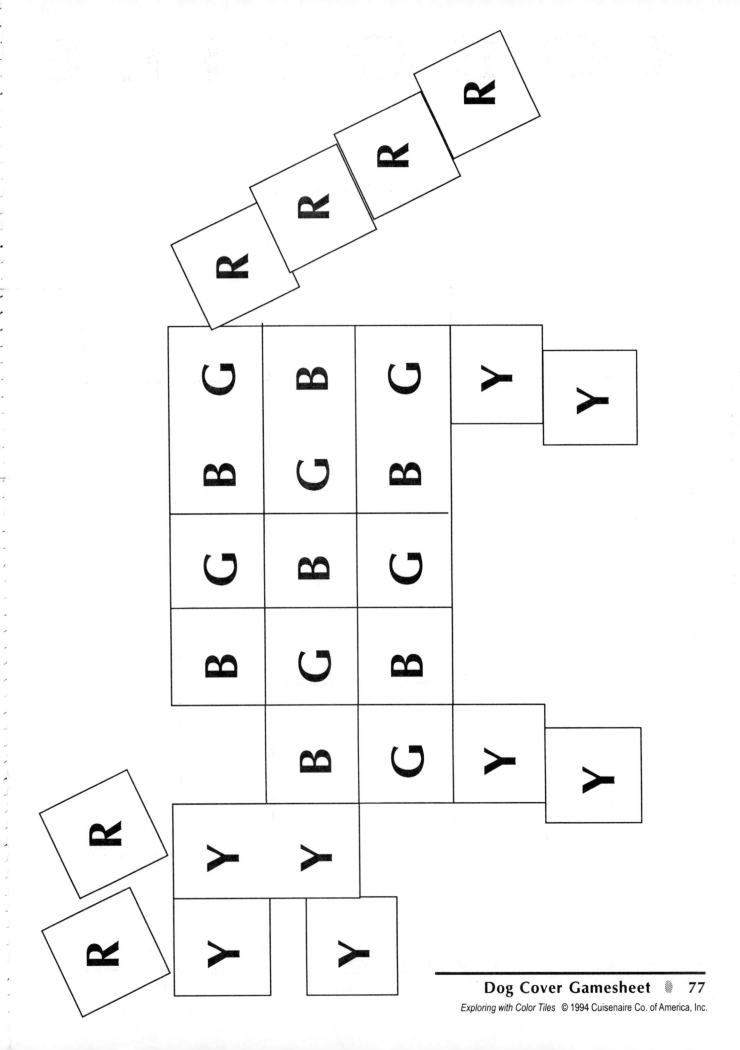

Exploring with Color Tiles © 1994 Cuisenaire Co. of America, Inc.

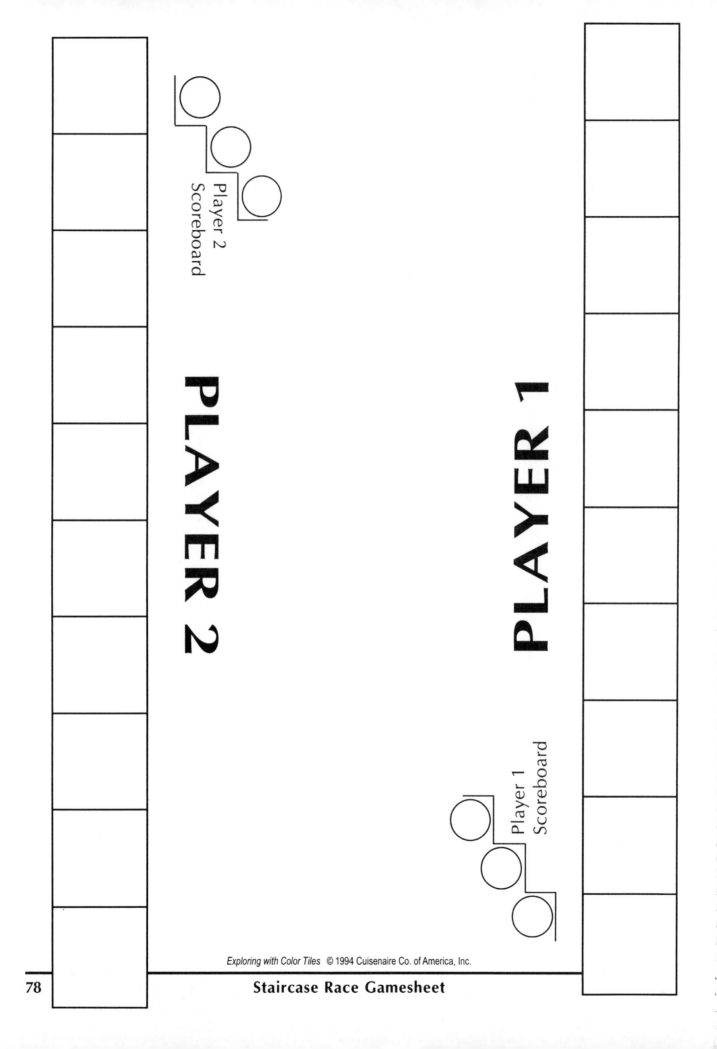

PLAYER 2

Player 2
Scoreboard

PLAYER 1

Player 1
Scoreboard

Exploring with Color Tiles © 1994 Cuisenaire Co. of America, Inc.

Staircase Race Gamesheet

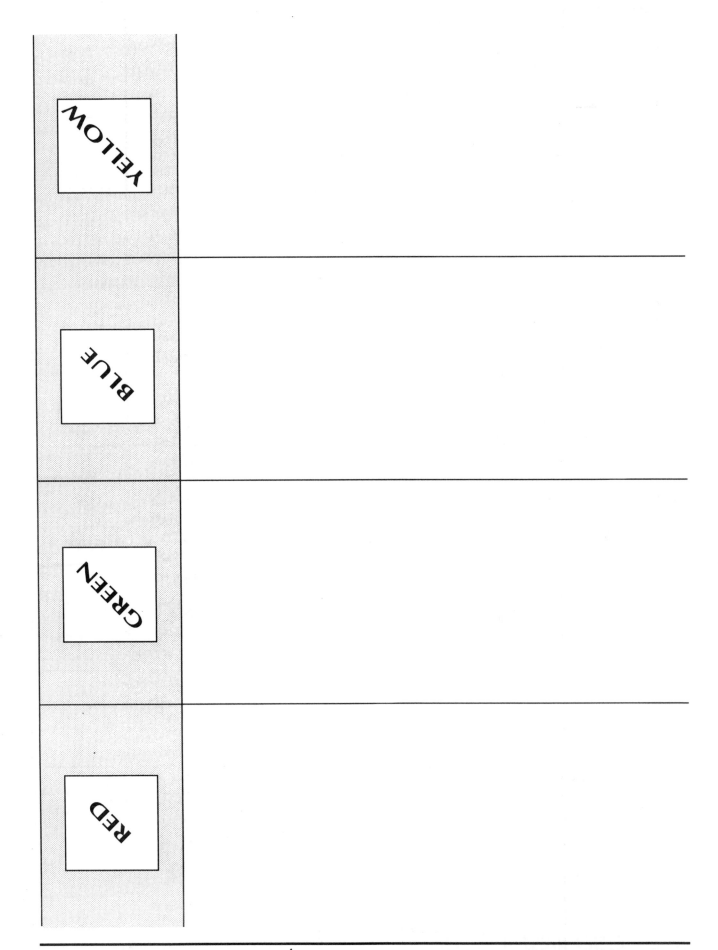

THE 100 CHART

1	2	3	4	5	6	7	8	9	10
11	12	13	14	15	16	17	18	19	20
21	22	23	24	25	26	27	28	29	30
31	32	33	34	35	36	37	38	39	40
41	42	43	44	45	46	47	48	49	50

CUT ON THIS SOLID LINE

51	52	53	54	55	56	57	58	59	60
61	62	63	64	65	66	67	68	69	70
71	72	73	74	75	76	77	78	79	80
81	82	83	84	85	86	87	88	89	90
91	92	93	94	95	96	97	98	99	100

NAME

NAME

The 100 Chart ❀ 81

Exploring with Color Tiles © 1994 Cuisenaire Co. of America, Inc.

3	5	7	1	9	8
4	10	2	9	6	5
8	2	6	7	1	2
10	7	3	4	10	8
5	6	4	1	3	9

SCORES

Player 1

NAME _____

1	2	3
☐	☐	☐

4	5	6	7
☐	☐	☐	☐

Total _____

Player 1

NAME _____

1	2	3
☐	☐	☐

4	5	6	7
☐	☐	☐	☐

Total _____

NAMES _____ _____

Guess _____

Count _____

Tens	Ones

NAMES _____ _____

Guess _____

Count _____

Tens	Ones

NAMES _____ _____   **TASK CARD C**

Guess _____

Count _____

Tens	Ones

NAMES _____ _____   **TASK CARD D**

Guess _____ Count _____

Tens	Ones

Exploring with Color Tiles © 1994 Cuisenaire Co. of America, Inc.

Guess _____

Count _____

Tens	Ones

Guess _____

Count _____

Tens	Ones

❋ **One-inch Squared Paper**

✸ **One-inch Squared Paper**

❋ **One-inch Squared Paper**